Fraction
Operations

Author   David Hudson
Editor   Kathy Rogers

Table of Contents

Equivalent Fractions Chart 2
Beginning Fractions Review 3
Beginning Fractions Review 4
Beginning Fractions Review 5
Adding Like Denominators 6
Subtracting Like Denominators 7
Common Multiples 8
Common Denominators 9
Adding Fractions 10
Adding Fractions 11
Subtracting Fractions 12
Subtracting Fractions 13
Multiplying Fractions 14
Multiplying Fractions 15
Multiplying Fractions 16
Dividing Fractions 17
Dividing Fractions 18
Review 1 19
Review 2 20
Fractions of Whole Numbers 21
Mixed Numbers 22
Mixed Numbers 23
Mixed Numbers 24
Practice 1 25
Practice 2 26
Assessment 1 27
Assessment 2 28

Answer Keys
Pages 3 - 8 29
Pages 9 - 15 30
Pages 16 - 22 31
Pages 23 - 28 32

Reproducible for classroom use only.
Not for use by an entire school or school system.

EP160 • ©1999, 2003 Edupress, Inc.™ • P.O. Box 883 • Dana Point, CA 92629
www.edupressinc.com
ISBN 1-56472-160-4
Printed in USA

Equivalent Fractions Chart

1								
$\frac{1}{2}$		$\frac{1}{4}$	$\frac{1}{8}$	$\frac{1}{12}$	$\frac{1}{3}$	$\frac{1}{6}$	$\frac{1}{9}$	

Fractions II—Operations
© Edupress EP160

Beginning Fractions Review

Advanced Fractions

Write the fractions for the shaded and unshaded parts of each shape or set.

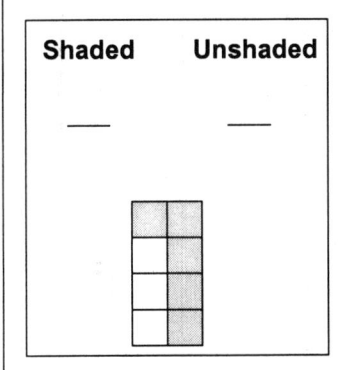

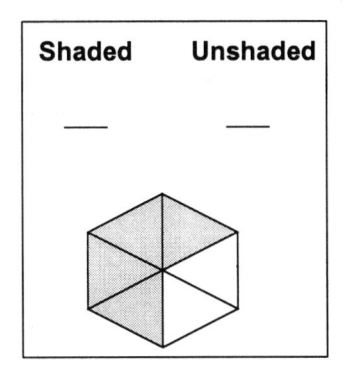

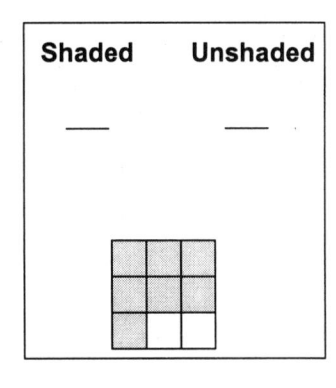

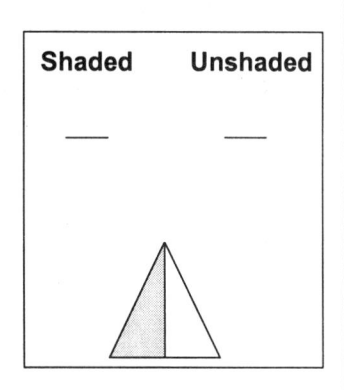

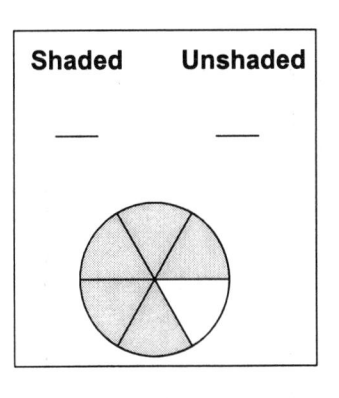

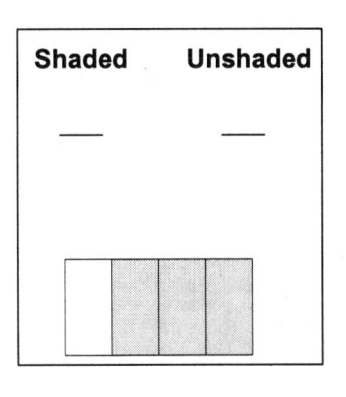

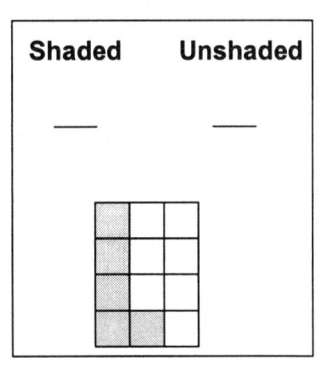

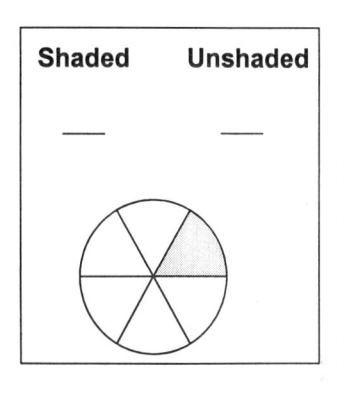

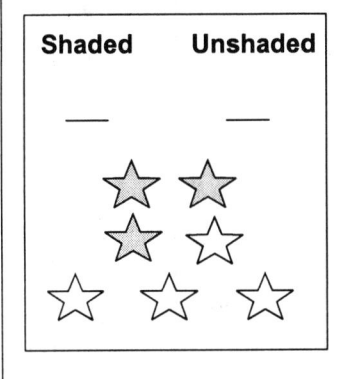

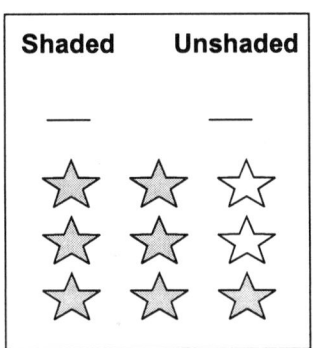

Shaded	Unshaded
___	___

Shaded	Unshaded
___	___

Shaded	Unshaded
___	___

Shaded	Unshaded
___	___

Shaded	Unshaded
___	___

Shaded	Unshaded
___	___

Shaded	Unshaded
___	___

Shaded	Unshaded
___	___

Shaded	Unshaded
___	___

Shaded	Unshaded
___	___

Shaded	Unshaded
___	___

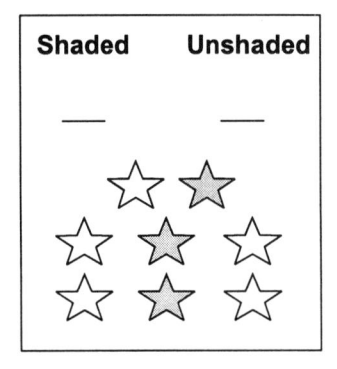

Shaded	Unshaded
___	___

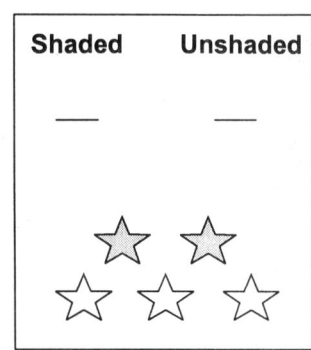

Write the fractional equivalent to one for each shape.

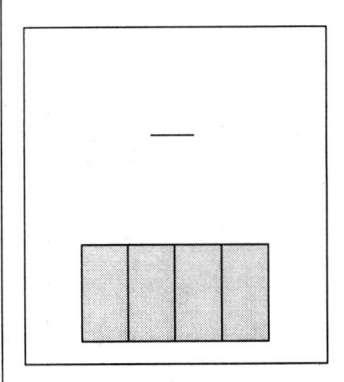

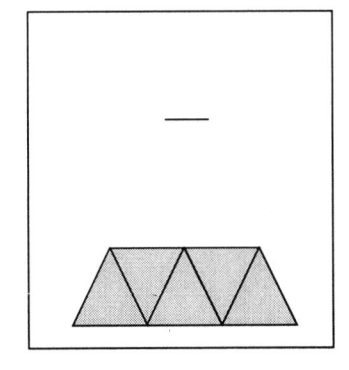

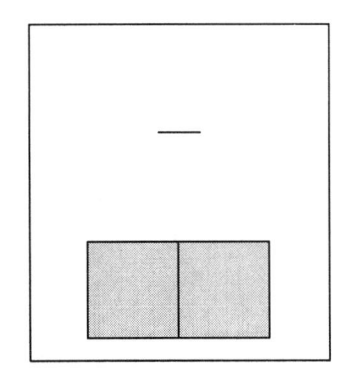

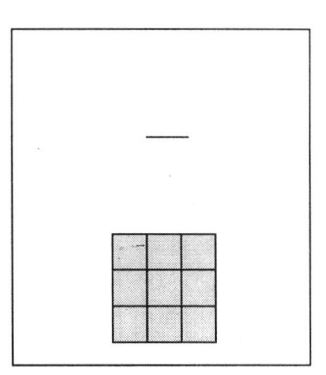

Fractions II—Operations

© Edupress EP160

Advanced Fractions

Beginning Fractions Review

Find the greatest common factor of each pair of numbers below and circle it.

60
18

35
21

36
15

Reduce these fractions to lowest terms.

$\dfrac{12}{18}$ $\dfrac{16}{32}$ $\dfrac{15}{25}$

$\dfrac{18}{21}$ $\dfrac{12}{27}$ $\dfrac{13}{39}$

Change these fractions to mixed numbers.

$\dfrac{24}{9}$ $\dfrac{7}{2}$ $\dfrac{14}{3}$

$\dfrac{36}{14}$ $\dfrac{10}{3}$ $\dfrac{11}{4}$

Fractions II—Operations © Edupress EP160

Advanced Fractions

Beginning Fractions Review

Change these mixed numbers to improper fractions.

$2\frac{3}{7}$	$3\frac{1}{4}$	$7\frac{1}{2}$
$1\frac{5}{9}$	$2\frac{3}{8}$	$5\frac{3}{5}$
$1\frac{7}{8}$	$1\frac{5}{6}$	$2\frac{2}{3}$

Make equivalent fractions.

$\dfrac{1}{5} = \dfrac{}{20}$ $\qquad$ $\dfrac{1}{7} = \dfrac{}{21}$ $\qquad$ $\dfrac{1}{8} = \dfrac{}{32}$

$\dfrac{3}{8} = \dfrac{}{16}$ $\qquad$ $\dfrac{4}{5} = \dfrac{}{30}$ $\qquad$ $\dfrac{3}{7} = \dfrac{}{28}$

$\dfrac{2}{9} = \dfrac{}{81}$ $\qquad$ $\dfrac{3}{7} = \dfrac{}{35}$ $\qquad$ $\dfrac{4}{9} = \dfrac{}{72}$

$\dfrac{3}{4} = \dfrac{}{24}$ $\qquad$ $\dfrac{5}{9} = \dfrac{}{18}$ $\qquad$ $\dfrac{7}{8} = \dfrac{}{16}$

$\dfrac{5}{32} = \dfrac{}{64}$ $\qquad$ $\dfrac{5}{6} = \dfrac{}{18}$ $\qquad$ $\dfrac{7}{11} = \dfrac{}{55}$

Fractions II—Operations $\qquad$ © Edupress EP160

Adding Like Denominators

Advanced Fractions

To add fractions with like denominators, add the numerators and keep the same denominator.

$$\frac{3}{6} + \frac{2}{6} = \frac{5}{6}$$

$$\frac{3}{8} + \frac{4}{8} = \frac{7}{8}$$

$\dfrac{1}{3} + \dfrac{1}{3} =$ $\dfrac{1}{4} + \dfrac{2}{4} =$ $\dfrac{3}{7} + \dfrac{2}{7} =$

$\dfrac{3}{10} + \dfrac{4}{10} =$ $\dfrac{2}{9} + \dfrac{6}{9} =$ $\dfrac{5}{12} + \dfrac{6}{12} =$

$\dfrac{2}{15} + \dfrac{5}{15} =$ $\dfrac{8}{18} + \dfrac{9}{18} =$ $\dfrac{6}{8} + \dfrac{1}{8} =$

$\dfrac{1}{9} + \dfrac{6}{9} =$ $\dfrac{2}{25} + \dfrac{9}{25} =$ $\dfrac{2}{16} + \dfrac{7}{16} =$

$\dfrac{1}{7} + \dfrac{5}{7} =$ $\dfrac{4}{11} + \dfrac{5}{11} =$ $\dfrac{5}{9} + \dfrac{2}{9} =$

Fractions II—Operations   © Edupress EP160

Subtracting Like Denominators
Advanced Fractions

To subtract fractions with like denominators, subtract the numerators and keep the same denominator.

$$\frac{5}{6} - \frac{4}{6} = \frac{1}{6}$$

$$\frac{6}{8} - \frac{3}{8} = \frac{3}{8}$$

$\frac{2}{3} - \frac{1}{3} =$ $\quad\quad\quad \frac{8}{9} - \frac{4}{9} =$ $\quad\quad\quad \frac{5}{7} - \frac{2}{7} =$

$\frac{3}{4} - \frac{2}{4} =$ $\quad\quad\quad \frac{6}{11} - \frac{2}{11} =$ $\quad\quad\quad \frac{7}{12} - \frac{2}{12} =$

$\frac{3}{5} - \frac{1}{5} =$ $\quad\quad\quad \frac{11}{18} - \frac{4}{18} =$ $\quad\quad\quad \frac{6}{8} - \frac{1}{8} =$

$\frac{7}{16} - \frac{2}{16} =$ $\quad\quad\quad \frac{9}{13} - \frac{2}{13} =$ $\quad\quad\quad \frac{7}{16} - \frac{2}{16} =$

$\frac{5}{7} - \frac{1}{7} =$ $\quad\quad\quad \frac{5}{8} - \frac{4}{8} =$ $\quad\quad\quad \frac{7}{9} - \frac{2}{9} =$

Fractions II—Operations © Edupress EP160

Advanced Fractions

Common Multiples

We can find multiples of numbers by multiplying them by 1, 2, 3, 4 and so on.

Multiples of 2: 2, 4, **6**, 8, 10, **12**, 14, 16, **18**, 20

Multiples of 3: 3, **6**, 9, **12**, 15, **18**

Common multiples are multiples that are the same for two or more numbers. The **lowest common multiple** is the smallest multiple of two or more numbers. 6 is the lowest common multiple of 2 and 3.

Find the lowest common multiple for each pair of numbers below and circle it.

Example

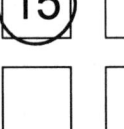

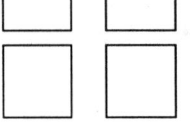

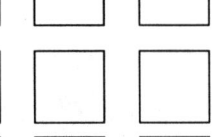

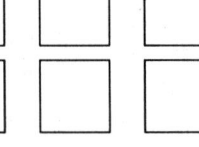

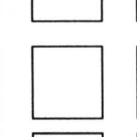

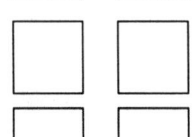

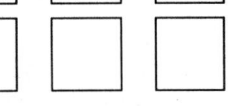

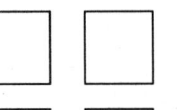

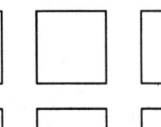

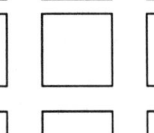

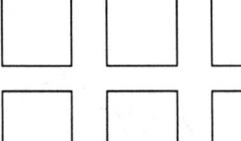

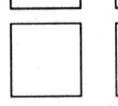

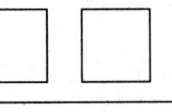

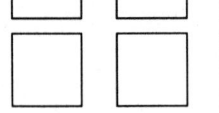

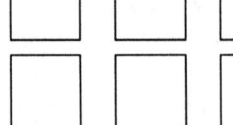

3	3	6	9	12	(15)
5	5	10	(15)		

3										
4										

4										
5										

6										
7										

6										
8										

7										
10										

Fractions II—Operations © Edupress EP160

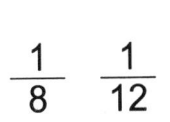

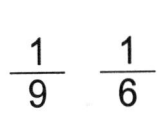

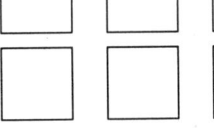

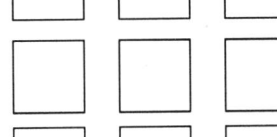

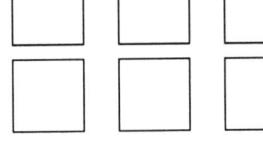

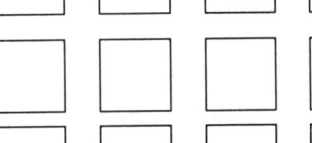

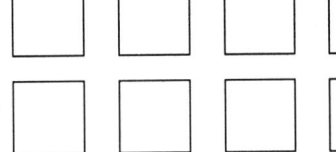

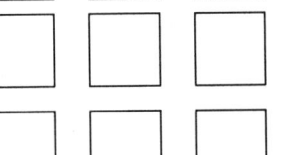

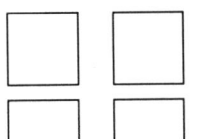

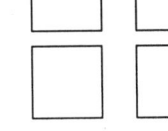

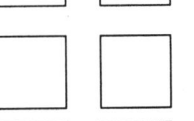

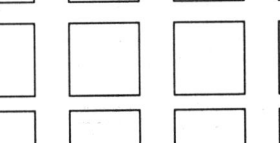

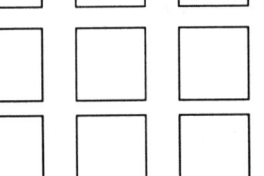

Advanced Fractions

Common Denominators

The **lowest common denominator** is the lowest common multiple of two or more denominators. You need to find the lowest common denominator when you are adding two fractions with different denominators.

Find the lowest common denominator for these pairs of fractions and circle it.

Example

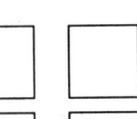

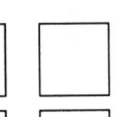

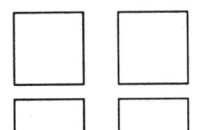

$\frac{1}{3}$ $\frac{1}{4}$ 3 6 9 ⑫
            4 8 ⑫

$\frac{1}{6}$ $\frac{1}{8}$

$\frac{1}{4}$ $\frac{1}{7}$

$\frac{1}{5}$ $\frac{1}{9}$

$\frac{1}{7}$ $\frac{1}{6}$

$\frac{1}{4}$ $\frac{1}{10}$

$\frac{1}{8}$ $\frac{1}{12}$

$\frac{1}{9}$ $\frac{1}{6}$

Fractions II—Operations

Advanced Fractions

Adding Fractions

To add fractions with different denominators:
1. Find a common denominator.
2. Convert the fractions to equivalent fractions with a common denominator.
3. Add the numerators.

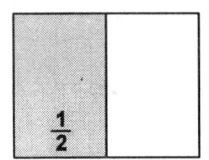

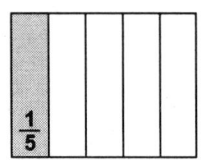

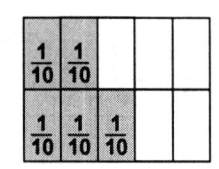

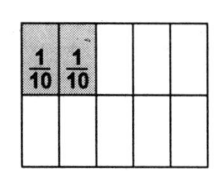

$$\frac{1}{2} + \frac{1}{5} = \frac{5}{10} + \frac{2}{10} = \frac{7}{10}$$

Common denominator = 10

Convert to tenths.   Convert to tenths.
$\frac{1 \times 5}{2 \times 5} = \frac{5}{10}$ $\frac{1 \times 2}{5 \times 2} = \frac{2}{10}$

Example

$\dfrac{3}{10} + \dfrac{2}{5} = \dfrac{3}{10} + \dfrac{4}{10} = \dfrac{7}{10}$

$\dfrac{2}{6} + \dfrac{1}{3} =$

$\dfrac{1}{6} + \dfrac{2}{12} =$

$\dfrac{1}{9} + \dfrac{1}{3} =$

$\dfrac{3}{8} + \dfrac{3}{16} =$

$\dfrac{2}{5} + \dfrac{2}{15} =$

$\dfrac{2}{3} + \dfrac{1}{18} =$

$\dfrac{1}{9} + \dfrac{2}{3} =$

$\dfrac{5}{9} + \dfrac{1}{3} =$

$\dfrac{4}{7} + \dfrac{5}{14} =$

$\dfrac{1}{6} + \dfrac{4}{12} =$

$\dfrac{1}{4} + \dfrac{1}{8} =$

Adding Fractions — Advanced Fractions

Reduce answers to lowest terms if necessary.

Example

$$\frac{1}{10} + \frac{2}{5} = \frac{1}{10} + \frac{4}{10} = \frac{5}{10} = \frac{1}{2}$$

Example

$$\frac{2}{3} + \frac{3}{4} = \frac{8}{12} + \frac{9}{12} = \frac{17}{12} = 1\frac{5}{12}$$

$\frac{1}{2} + \frac{1}{4} =$

$\frac{1}{8} + \frac{1}{2} =$

$\frac{3}{8} + \frac{1}{4} =$

$\frac{1}{2} + \frac{3}{5} =$

$\frac{5}{12} + \frac{3}{4} =$

$\frac{5}{6} + \frac{2}{7} =$

$\frac{1}{3} + \frac{2}{7} =$

$\frac{1}{9} + \frac{3}{8} =$

$\frac{4}{7} + \frac{7}{10} =$

$\frac{7}{8} + \frac{1}{3} =$

$\frac{3}{5} + \frac{2}{25} =$

$\frac{1}{7} + \frac{5}{8} =$

$\frac{1}{15} + \frac{3}{10} =$

$\frac{3}{5} + \frac{5}{7} =$

$\frac{1}{4} + \frac{2}{3} =$

$\frac{3}{10} + \frac{2}{5} =$

$\frac{3}{8} + \frac{3}{10} =$

$\frac{5}{9} + \frac{4}{7} =$

Fractions II—Operations

Subtracting Fractions

Advanced Fractions

To subtract fractions with different denominators:
1. Find a common denominator.
2. Convert the fractions to equivalent fractions with a common denominator.
3. Subtract the numerators.

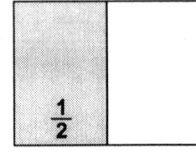

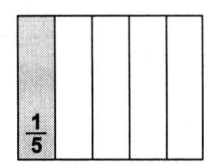

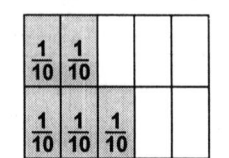

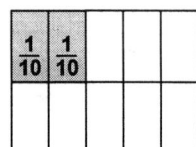

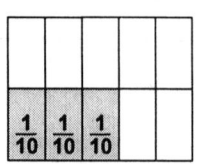

Common denominator = 10

$$\frac{1}{2} - \frac{1}{5} = \frac{5}{10} - \frac{2}{10} = \frac{3}{10}$$

Convert to tenths. Convert to tenths.
$\frac{1 \times 5}{2 \times 5} = \frac{5}{10}$ $\frac{1 \times 2}{5 \times 2} = \frac{2}{10}$

Example

$\frac{1}{2} - \frac{1}{3} = \frac{3}{6} - \frac{2}{6} = \frac{1}{6}$

$\frac{2}{5} - \frac{2}{15} =$

$\frac{3}{4} - \frac{1}{3} =$

$\frac{1}{6} - \frac{1}{8} =$

$\frac{5}{9} - \frac{1}{3} =$

$\frac{7}{15} - \frac{1}{5} =$

$\frac{3}{5} - \frac{1}{4} =$

$\frac{3}{4} - \frac{1}{6} =$

$\frac{7}{8} - \frac{1}{2} =$

$\frac{5}{6} - \frac{2}{3} =$

$\frac{8}{9} - \frac{1}{3} =$

$\frac{7}{9} - \frac{7}{18} =$

Fractions II—Operations © Edupress EP160

Subtracting Fractions — Advanced Fractions

Reduce answers to lowest terms if necessary.

Example

$$\frac{6}{10} - \frac{2}{5} = \frac{6}{10} - \frac{4}{10} = \frac{2}{10} = \frac{1}{5}$$

Example

$$\frac{3}{4} - \frac{2}{3} = \frac{9}{12} - \frac{8}{12} = \frac{1}{12}$$

$\frac{7}{8} - \frac{3}{16} =$

$\frac{1}{2} - \frac{1}{8} =$

$\frac{3}{8} - \frac{1}{6} =$

$\frac{3}{5} - \frac{1}{2} =$

$\frac{11}{12} - \frac{3}{4} =$

$\frac{9}{13} - \frac{5}{26} =$

$\frac{5}{9} - \frac{1}{3} =$

$\frac{1}{9} - \frac{1}{54} =$

$\frac{6}{7} - \frac{7}{10} =$

$\frac{7}{8} - \frac{1}{3} =$

$\frac{3}{5} - \frac{2}{25} =$

$\frac{5}{8} - \frac{1}{7} =$

$\frac{5}{17} - \frac{3}{34} =$

$\frac{5}{7} - \frac{3}{5} =$

$\frac{1}{4} - \frac{3}{32} =$

$\frac{7}{10} - \frac{3}{5} =$

$\frac{3}{8} - \frac{3}{10} =$

$\frac{7}{9} - \frac{4}{7} =$

Fractions II—Operations

Multiplying Fractions

Advanced Fractions

To multiply fractions, multiply the numerator by the numerator and the denominator by the denominator.

$\dfrac{1}{2} \times \dfrac{1}{2} = \dfrac{1}{4}$ is the same as $\dfrac{1}{2}$ of $\dfrac{1}{2} = \dfrac{1}{4}$

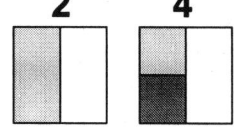

$\dfrac{2}{3} \times \dfrac{3}{4} = \dfrac{6}{12} = \dfrac{1}{2}$ is the same as $\dfrac{2}{3}$ of $\dfrac{3}{4} = \dfrac{1}{2}$

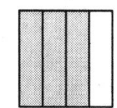

Example

$\dfrac{1}{3} \times \dfrac{1}{2} = \dfrac{1}{6}$

$\dfrac{1}{4} \times \dfrac{2}{3} =$

$\dfrac{5}{6} \times \dfrac{1}{4} =$

$\dfrac{3}{7} \times \dfrac{1}{2} =$

$\dfrac{4}{9} \times \dfrac{1}{3} =$

$\dfrac{5}{12} \times \dfrac{2}{3} =$

Example

$\dfrac{3}{5} \times \dfrac{1}{3} = \dfrac{3}{15} = \dfrac{1}{5}$

$\dfrac{2}{3} \times \dfrac{1}{3} =$

$\dfrac{4}{9} \times \dfrac{3}{4} =$

$\dfrac{5}{8} \times \dfrac{4}{5} =$

$\dfrac{3}{7} \times \dfrac{1}{4} =$

$\dfrac{1}{6} \times \dfrac{1}{9} =$

Fractions II—Operations

Advanced Fractions

Multiplying Fractions

Canceling makes multiplying fractions easier. Divide the numerator of one fraction and the denominator of the other by their greatest common factor.

The greatest common factor of 4 and 16 is 4. Divide both 4 and 16 by 4.

$$\frac{\cancel{4}^1}{5} \times \frac{3}{\cancel{16}_4} = \frac{1}{5} \times \frac{3}{4} = \frac{3}{20}$$

The greatest common factor of 8 and 12 is 4. Divide both 8 and 12 by 4.

The greatest common factor of 5 and 25 is 5. Divide both 5 and 25 by 5.

$$\frac{\cancel{8}^2}{\cancel{25}_5} \times \frac{\cancel{5}^1}{\cancel{12}_3} = \frac{2}{5} \times \frac{1}{3} = \frac{2}{15}$$

Cancel these fractions and multiply. Reduce answers to lowest terms if needed.

Example

$$\frac{5}{\cancel{6}_3} \times \frac{\cancel{2}^1}{3} = \frac{5}{9}$$

Example

$$\frac{\cancel{7}^1}{\cancel{33}_3} \times \frac{\cancel{11}^1}{\cancel{21}_3} = \frac{1}{9}$$

$$\frac{3}{4} \times \frac{7}{9} =$$

$$\frac{2}{15} \times \frac{5}{8} =$$

$$\frac{6}{49} \times \frac{7}{9} =$$

$$\frac{4}{9} \times \frac{3}{24} =$$

$$\frac{3}{7} \times \frac{7}{12} =$$

$$\frac{5}{8} \times \frac{4}{5} =$$

$$\frac{12}{21} \times \frac{7}{36} =$$

$$\frac{3}{10} \times \frac{2}{3} =$$

$$\frac{5}{12} \times \frac{9}{20} =$$

$$\frac{8}{81} \times \frac{9}{16} =$$

Fractions II—Operations

Advanced Fractions

Multiplying Fractions

Reduce the answers to lowest terms if needed.

$\dfrac{2}{3} \times \dfrac{3}{4} =$ $\qquad$ $\dfrac{1}{8} \times \dfrac{1}{9} =$

$\dfrac{5}{6} \times \dfrac{7}{15} =$ $\qquad$ $\dfrac{4}{9} \times \dfrac{3}{4} =$

$\dfrac{3}{5} \times \dfrac{5}{12} =$ $\qquad$ $\dfrac{6}{7} \times \dfrac{5}{9} =$

$\dfrac{30}{51} \times \dfrac{17}{20} =$ $\qquad$ $\dfrac{21}{22} \times \dfrac{11}{21} =$

$\dfrac{15}{16} \times \dfrac{3}{8} =$ $\qquad$ $\dfrac{2}{7} \times \dfrac{7}{8} =$

$\dfrac{5}{11} \times \dfrac{44}{45} =$ $\qquad$ $\dfrac{3}{5} \times \dfrac{8}{9} =$

$\dfrac{5}{24} \times \dfrac{8}{15} =$ $\qquad$ $\dfrac{11}{28} \times \dfrac{11}{15} =$

$\dfrac{7}{12} \times \dfrac{3}{7} =$ $\qquad$ $\dfrac{5}{21} \times \dfrac{3}{5} =$

$\dfrac{5}{6} \times \dfrac{3}{10} =$ $\qquad$ $\dfrac{9}{15} \times \dfrac{5}{27} =$

$\dfrac{8}{25} \times \dfrac{5}{24} =$ $\qquad$ $\dfrac{3}{7} \times \dfrac{2}{3} =$

Dividing Fractions — Advanced Fractions

To divide fractions, invert the divisor and multiply.

To find how many one fourths are in one half, divide one half by one fourth.

$$\frac{1}{2} \div \frac{1}{4} = \frac{1}{2} \times \frac{4}{1} = \frac{4}{2} = 2$$

There are 2 $\frac{1}{4}$'s in $\frac{1}{2}$

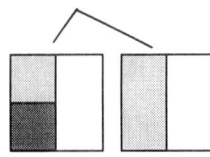

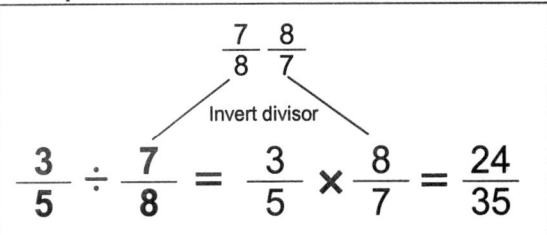

Example

$$\frac{3}{5} \div \frac{7}{8} = \frac{3}{5} \times \frac{8}{7} = \frac{24}{35}$$

Example

$$\frac{7}{8} \div \frac{1}{3} = \frac{7}{8} \times \frac{3}{1} = \frac{21}{8} = 2\frac{5}{8}$$

$\dfrac{1}{4} \div \dfrac{1}{3} =$

$\dfrac{1}{3} \div \dfrac{1}{4} =$

$\dfrac{2}{5} \div \dfrac{2}{3} =$

$\dfrac{1}{8} \div \dfrac{3}{4} =$

$\dfrac{5}{6} \div \dfrac{1}{8} =$

$\dfrac{2}{7} \div \dfrac{5}{6} =$

$\dfrac{2}{7} \div \dfrac{1}{14} =$

$\dfrac{1}{9} \div \dfrac{4}{11} =$

$\dfrac{1}{12} \div \dfrac{1}{36} =$

$\dfrac{7}{8} \div \dfrac{2}{3} =$

Fractions II—Operations © Edupress EP160

Advanced Fractions

Dividing Fractions

Reduce answers to lowest terms if needed.

$\dfrac{2}{3} \div \dfrac{3}{4} =$ $\qquad\qquad$ $\dfrac{3}{8} \div \dfrac{2}{5} =$

$\dfrac{4}{7} \div \dfrac{3}{7} =$ $\qquad\qquad$ $\dfrac{4}{9} \div \dfrac{3}{4} =$

$\dfrac{3}{5} \div \dfrac{5}{12} =$ $\qquad\qquad$ $\dfrac{8}{15} \div \dfrac{5}{9} =$

$\dfrac{5}{9} \div \dfrac{3}{4} =$ $\qquad\qquad$ $\dfrac{11}{14} \div \dfrac{11}{21} =$

$\dfrac{15}{16} \div \dfrac{8}{9} =$ $\qquad\qquad$ $\dfrac{2}{3} \div \dfrac{5}{8} =$

$\dfrac{5}{12} \div \dfrac{7}{12} =$ $\qquad\qquad$ $\dfrac{3}{8} \div \dfrac{8}{9} =$

$\dfrac{5}{24} \div \dfrac{5}{8} =$ $\qquad\qquad$ $\dfrac{4}{7} \div \dfrac{8}{9} =$

$\dfrac{7}{12} \div \dfrac{3}{7} =$ $\qquad\qquad$ $\dfrac{6}{7} \div \dfrac{5}{9} =$

$\dfrac{5}{6} \div \dfrac{3}{7} =$ $\qquad\qquad$ $\dfrac{8}{27} \div \dfrac{4}{9} =$

$\dfrac{8}{25} \div \dfrac{4}{5} =$ $\qquad\qquad$ $\dfrac{3}{7} \div \dfrac{2}{3} =$

Fractions II—Operations © Edupress EP160

Review 1 — Advanced Fractions

Reduce answers to lowest terms if needed.

$\dfrac{1}{8} + \dfrac{5}{12} =$ $\dfrac{3}{5} \times \dfrac{5}{8} =$

$\dfrac{4}{9} - \dfrac{1}{3} =$ $\dfrac{5}{9} \div \dfrac{4}{5} =$

$\dfrac{3}{8} \times \dfrac{8}{9} =$ $\dfrac{7}{15} + \dfrac{2}{3} =$

$\dfrac{3}{4} \div \dfrac{5}{9} =$ $\dfrac{3}{7} - \dfrac{8}{21} =$

$\dfrac{5}{16} + \dfrac{3}{8} =$ $\dfrac{5}{8} \div \dfrac{2}{3} =$

$\dfrac{9}{10} - \dfrac{5}{6} =$ $\dfrac{3}{8} \times \dfrac{8}{9} =$

$\dfrac{3}{7} \div \dfrac{7}{8} =$ $\dfrac{8}{9} + \dfrac{4}{7} =$

$\dfrac{7}{12} \times \dfrac{3}{7} =$ $\dfrac{6}{7} - \dfrac{1}{3} =$

$\dfrac{5}{6} - \dfrac{3}{7} =$ $\dfrac{5}{6} \div \dfrac{1}{6} =$

$\dfrac{8}{15} + \dfrac{2}{3} =$ $\dfrac{3}{7} \times \dfrac{2}{3} =$

Review 2

Advanced Fractions

Reduce answers to lowest terms if needed.

$\dfrac{1}{3} + \dfrac{4}{5} =$ $\dfrac{3}{7} \times \dfrac{2}{3} =$

$\dfrac{7}{8} - \dfrac{1}{5} =$ $\dfrac{3}{8} \div \dfrac{3}{4} =$

$\dfrac{3}{7} \times \dfrac{14}{15} =$ $\dfrac{6}{11} + \dfrac{5}{6} =$

$\dfrac{4}{9} \div \dfrac{5}{9} =$ $\dfrac{3}{13} - \dfrac{3}{26} =$

$\dfrac{5}{8} + \dfrac{7}{12} =$ $\dfrac{5}{8} \div \dfrac{5}{6} =$

$\dfrac{4}{7} - \dfrac{5}{9} =$ $\dfrac{11}{12} \times \dfrac{8}{9} =$

$\dfrac{7}{8} \div \dfrac{6}{7} =$ $\dfrac{1}{9} + \dfrac{3}{8} =$

$\dfrac{7}{9} \times \dfrac{1}{7} =$ $\dfrac{5}{11} - \dfrac{1}{6} =$

$\dfrac{3}{10} - \dfrac{1}{7} =$ $\dfrac{5}{6} \div \dfrac{2}{5} =$

$\dfrac{8}{9} + \dfrac{2}{3} =$ $\dfrac{6}{7} \times \dfrac{7}{8} =$

Fractions II—Operations

© Edupress EP160

Fractions of Whole Numbers — Advanced Fractions

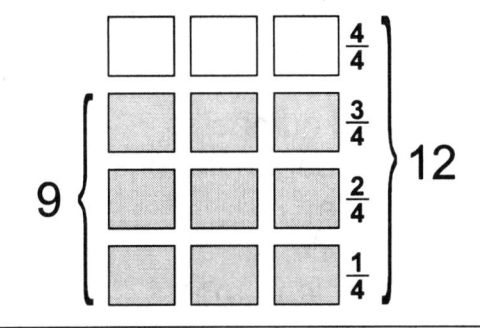

To find a fraction of a whole number:

Convert the whole number to fractional form and multiply.

$$\frac{3}{4} \text{ of } 12 = \frac{3}{4} \times \frac{12}{1} = 9$$

or

Divide the whole number by the denominator and then multiply by the numerator.

$$\frac{3}{4} \text{ of } 12 = 4\overline{)12} = 3 \quad 3 \times 3 = 9$$

$$\frac{3}{4} \text{ of } 12$$

Example

$$\frac{2}{3} \times 9 = \frac{2}{3} \times \frac{9}{1} = \frac{18}{3} = 6$$

$$\frac{1}{5} \times 30 =$$

$$\frac{5}{6} \times 48 =$$

$$\frac{3}{7} \times 42 =$$

$$\frac{2}{11} \times 15 =$$

$$\frac{3}{4} \times 7 =$$

Example

$$\frac{3}{4} \times 9 = \frac{3}{4} \times \frac{9}{1} = \frac{27}{4} = 6\frac{3}{4}$$

$$\frac{1}{8} \times 32 =$$

$$\frac{2}{9} \times 25 =$$

$$\frac{5}{8} \times 7 =$$

$$\frac{1}{16} \times 32 =$$

$$\frac{3}{50} \times 50 =$$

Fractions II—Operations

Mixed Numbers — Advanced Fractions

To add mixed numbers:
  Find a common denominator.
  Add the whole numbers.
  Add the fractions.
  Reduce to lowest terms.

$$9\tfrac{2}{3} \quad = \quad 9\tfrac{8}{12}$$
$$+\,7\tfrac{3}{4} \quad\quad +\,7\tfrac{9}{12}$$
$$\overline{}$$
$$16\tfrac{17}{12} \;=\; 17\tfrac{5}{12}$$

(Find a common denominator.)

To subtract mixed numbers:
  Find a common denominator.
  Regroup if needed.
  Subtract the fractions.
  Subtract the whole numbers.

$$6\tfrac{1}{3} \;=\; 6\tfrac{2}{6} \;=\; 5\tfrac{8}{6}$$
$$-\,3\tfrac{1}{2} \quad -\,3\tfrac{3}{6} \quad -\,3\tfrac{3}{6}$$
$$\overline{}$$
$$2\tfrac{5}{6}$$

(Find a common denominator.   Regroup)

$4\tfrac{3}{5}$ $2\tfrac{1}{3}$ $6\tfrac{8}{9}$
$+\,2\tfrac{3}{4}$ $+\,5\tfrac{1}{5}$ $+\,4\tfrac{7}{9}$

$5\tfrac{1}{4}$ $8\tfrac{1}{7}$ $4\tfrac{2}{3}$
$-\,4\tfrac{3}{4}$ $-\,2\tfrac{5}{6}$ $-\,2\tfrac{1}{5}$

Fractions II—Operations

Mixed Numbers — Advanced Fractions

To multiply mixed numbers, convert to improper fractions and multiply.

$$2\tfrac{5}{8} \times 3\tfrac{1}{3} = \tfrac{21}{8} \times \tfrac{10}{3} = \tfrac{21}{8} \times \tfrac{10}{3} = \tfrac{35}{4} = 8\tfrac{3}{4}$$

To divide mixed numbers, convert to improper fractions, invert the divisor and multiply.

$$3\tfrac{1}{4} \div 2\tfrac{1}{5} = \tfrac{13}{4} \div \tfrac{11}{5} = \tfrac{13}{4} \times \tfrac{5}{11} = \tfrac{65}{44} = 1\tfrac{21}{44}$$

$3\tfrac{3}{4} \times 2\tfrac{1}{5} =$

$5\tfrac{2}{3} \div 2\tfrac{1}{2} =$

$8\tfrac{5}{12} \times 3\tfrac{1}{3} =$

$5\tfrac{2}{7} \div 2\tfrac{7}{8} =$

$6\tfrac{5}{8} \times 3\tfrac{2}{3} =$

$2\tfrac{3}{4} \div 2\tfrac{1}{12} =$

$4\tfrac{1}{6} \times 2\tfrac{2}{7} =$

$3\tfrac{1}{9} \div 3\tfrac{1}{3} =$

Mixed Numbers — **Advanced Fractions**

Reduce answers to lowest terms if needed.

$2 \frac{2}{5} + 3 \frac{3}{5} =$

$9 \frac{1}{3} - 7 \frac{2}{3} =$

$3 \frac{1}{2} \div 2 \frac{1}{7} =$

$2 \frac{3}{8} \times 4 \frac{2}{7} =$

$8 \frac{2}{5} - 6 \frac{1}{2} =$

$9 \frac{3}{11} + 2 \frac{1}{12} =$

$3 \frac{1}{7} \div 2 \frac{3}{5} =$

$2 \frac{5}{8} \times 3 \frac{1}{3} =$

$7 \frac{3}{4} + 2 \frac{5}{12} =$

$6 \frac{3}{5} - 3 \frac{1}{7} =$

$9 \frac{1}{6} \div 3 \frac{1}{3} =$

Fractions II—Operations © Edupress EP160

Practice 1

Advanced Fractions

$\dfrac{3}{7} + \dfrac{3}{4} =$ $\dfrac{5}{6} \times \dfrac{6}{7} =$

$\dfrac{8}{15} - \dfrac{1}{3} =$ $\dfrac{7}{8} \div \dfrac{7}{9} =$

$\dfrac{3}{5} \times \dfrac{2}{7} =$ $\dfrac{4}{9} + \dfrac{5}{12} =$

$\dfrac{3}{7} \div \dfrac{3}{4} =$ $\dfrac{3}{5} - \dfrac{2}{25} =$

$\dfrac{5}{7} + \dfrac{3}{8} =$ $\dfrac{5}{9} \div \dfrac{2}{3} =$

$\dfrac{5}{6} - \dfrac{7}{10} =$ $\dfrac{11}{15} \times \dfrac{5}{22} =$

$7\dfrac{5}{12} + 2\dfrac{3}{4} =$

$1\dfrac{2}{5} \times 1\dfrac{1}{3} =$

$3\dfrac{3}{7} + 5\dfrac{5}{14} =$

$6 - 3\dfrac{1}{7} =$

$7 \div 1\dfrac{4}{9} =$

Fractions II—Operations   © Edupress EP160

Practice 2 — Advanced Fractions

$\dfrac{3}{5} + \dfrac{2}{3} =$ $\dfrac{9}{10} \times \dfrac{2}{5} =$

$\dfrac{6}{7} - \dfrac{1}{5} =$ $\dfrac{4}{9} \div \dfrac{5}{9} =$

$\dfrac{9}{15} \times \dfrac{3}{7} =$ $\dfrac{13}{15} + \dfrac{5}{12} =$

$\dfrac{21}{22} \div \dfrac{2}{7} =$ $\dfrac{4}{5} - \dfrac{2}{7} =$

$\dfrac{3}{11} + \dfrac{2}{10} =$ $\dfrac{6}{7} \div \dfrac{1}{3} =$

$\dfrac{4}{5} - \dfrac{4}{7} =$ $\dfrac{12}{13} \times \dfrac{5}{17} =$

$4\dfrac{2}{7} + 5\dfrac{3}{5} =$

$3\dfrac{2}{9} \times 4\dfrac{5}{6} =$

$2\dfrac{3}{8} + 7\dfrac{1}{6} =$

$8 - 1\dfrac{3}{5} =$

$6 \div 2\dfrac{2}{5} =$

Fractions II—Operations

Assessment 1 — Advanced Fractions

Reduce answers to lowest terms if needed.

$\frac{2}{7} + \frac{3}{7} =$ $\qquad$ $\frac{3}{8} + \frac{3}{8} =$

$\frac{4}{9} - \frac{2}{9} =$ $\qquad$ $\frac{7}{10} - \frac{3}{10} =$

$\frac{5}{8} + \frac{4}{9} =$ $\qquad$ $\frac{3}{8} + \frac{5}{6} =$

$\frac{3}{25} + \frac{7}{20} =$ $\qquad$ $\frac{3}{16} + \frac{5}{6} =$

$\frac{13}{16} - \frac{5}{8} =$ $\qquad$ $\frac{25}{72} - \frac{1}{8} =$

$\frac{1}{5} \times \frac{2}{3} =$ $\qquad$ $\frac{3}{4} \times \frac{5}{7} =$

$\frac{9}{14} \times \frac{21}{27} =$ $\qquad$ $\frac{8}{9} \times \frac{4}{7} =$

$\frac{6}{7} \times \frac{21}{29} =$ $\qquad$ $\frac{12}{13} \times \frac{65}{144} =$

$\frac{2}{3} \div \frac{1}{3} =$ $\qquad$ $\frac{1}{4} \div \frac{1}{2} =$

$\frac{14}{15} \div \frac{7}{15} =$ $\qquad$ $\frac{7}{51} \div \frac{14}{17} =$

Assessment 2 **Advanced Fractions**

Reduce answers to lowest terms if needed.

$\frac{2}{5} \times 9 =$

$\frac{11}{12} \times 16 =$

$7 - \frac{1}{7} =$

$8 - \frac{11}{13} =$

$5\frac{8}{9} + 4\frac{1}{2} =$

$4\frac{4}{15} + 3\frac{5}{12} =$

$6\frac{1}{4} - 4\frac{1}{3} =$

$7\frac{5}{8} \times 3\frac{1}{3} =$

$7\frac{3}{4} \times 2\frac{5}{12} =$

$6\frac{3}{5} \div 3\frac{1}{7} =$

$9\frac{1}{6} \div 3\frac{1}{3} =$

Fractions II—Operations

Advanced Fractions

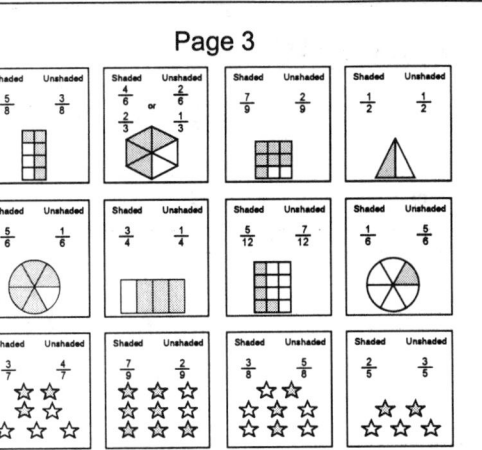

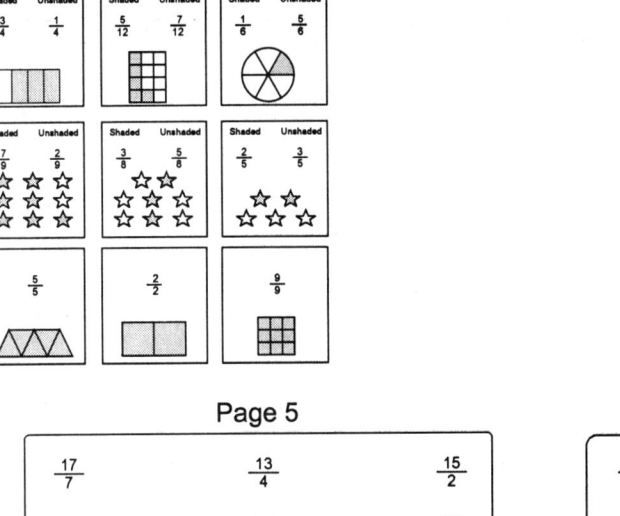

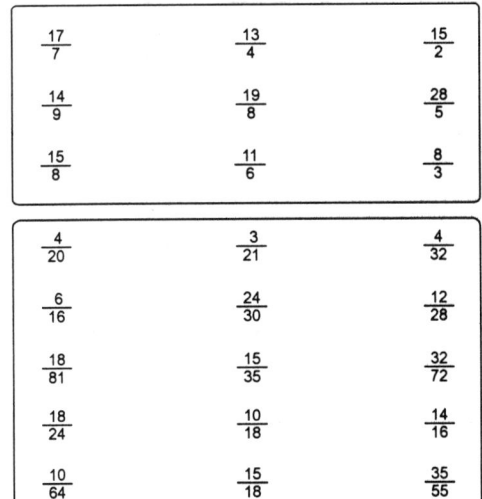

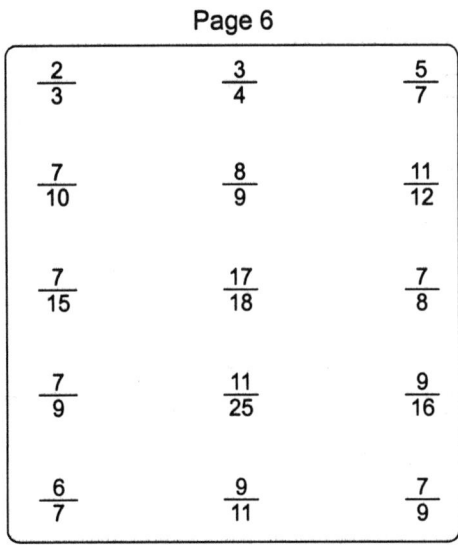

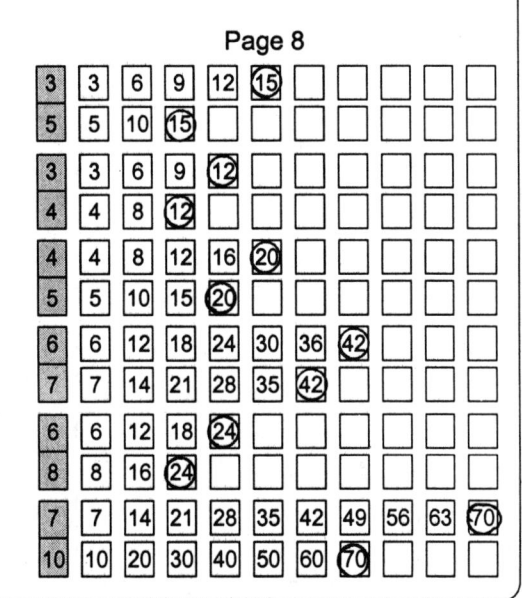

Answers Pages 3 - 8

Page 3

Shaded	Unshaded
$\frac{5}{8}$	$\frac{3}{8}$

Shaded	Unshaded
$\frac{4}{6}$ or $\frac{2}{3}$	$\frac{2}{6}$ or $\frac{1}{3}$

Shaded	Unshaded
$\frac{7}{9}$	$\frac{2}{9}$

Shaded	Unshaded
$\frac{1}{2}$	$\frac{1}{2}$

Shaded	Unshaded
$\frac{5}{6}$	$\frac{1}{6}$

Shaded	Unshaded
$\frac{3}{4}$	$\frac{1}{4}$

Shaded	Unshaded
$\frac{5}{12}$	$\frac{7}{12}$

Shaded	Unshaded
$\frac{1}{6}$	$\frac{5}{6}$

Shaded	Unshaded
$\frac{3}{7}$	$\frac{4}{7}$

Shaded	Unshaded
$\frac{7}{9}$	$\frac{2}{9}$

Shaded	Unshaded
$\frac{5}{8}$	$\frac{3}{8}$

Shaded	Unshaded
$\frac{2}{5}$	$\frac{3}{5}$

$\frac{4}{4}$ $\frac{5}{5}$ $\frac{2}{2}$ $\frac{9}{9}$

Page 4

60	1	2	3	4	5	⑥
18	1	2	3	⑥
35	1	5	⑦
21	1	3	⑦
36	1	2	③
15	1	③

$\frac{2}{3}$ $\frac{1}{2}$ $\frac{3}{5}$

$\frac{6}{7}$ $\frac{4}{9}$ $\frac{1}{3}$

$2\frac{2}{3}$ $3\frac{1}{2}$ $4\frac{2}{3}$

$2\frac{4}{7}$ $3\frac{1}{3}$ $2\frac{3}{4}$

Page 5

$\frac{17}{7}$ $\frac{13}{4}$ $\frac{15}{2}$

$\frac{14}{9}$ $\frac{19}{8}$ $\frac{28}{5}$

$\frac{15}{8}$ $\frac{11}{6}$ $\frac{8}{3}$

$\frac{4}{20}$ $\frac{3}{21}$ $\frac{4}{32}$

$\frac{6}{16}$ $\frac{24}{30}$ $\frac{12}{28}$

$\frac{18}{81}$ $\frac{15}{35}$ $\frac{32}{72}$

$\frac{18}{24}$ $\frac{10}{18}$ $\frac{14}{16}$

$\frac{10}{64}$ $\frac{15}{18}$ $\frac{35}{55}$

Page 6

$\frac{2}{3}$ $\frac{3}{4}$ $\frac{5}{7}$

$\frac{7}{10}$ $\frac{8}{9}$ $\frac{11}{12}$

$\frac{7}{15}$ $\frac{17}{18}$ $\frac{7}{8}$

$\frac{7}{9}$ $\frac{11}{25}$ $\frac{9}{16}$

$\frac{6}{7}$ $\frac{9}{11}$ $\frac{7}{9}$

Page 7

$\frac{1}{3}$ $\frac{4}{9}$ $\frac{3}{7}$

$\frac{1}{4}$ $\frac{4}{11}$ $\frac{5}{12}$

$\frac{2}{5}$ $\frac{7}{18}$ $\frac{5}{8}$

$\frac{5}{16}$ $\frac{7}{13}$ $\frac{5}{16}$

$\frac{4}{7}$ $\frac{1}{8}$ $\frac{5}{9}$

Page 8

3	3	6	9	12	⑮
5	5	10	⑮
3	3	6	9	⑫
4	4	8	⑫
4	4	8	12	16	⑳
5	5	10	15	⑳
6	6	12	18	24	30	36	㊷
7	7	14	21	28	35	㊷
6	6	12	18	㉔
8	8	16	㉔
7	7	14	21	28	35	42	49	56	63	㊀
10	10	20	30	40	50	60	㊀

Fractions II—Operations   © Edupress EP160

Advanced Fractions

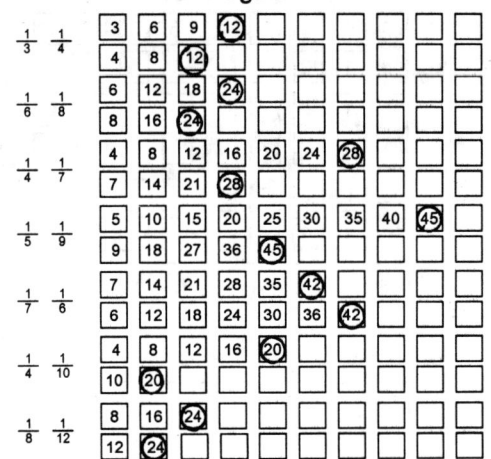

Answers Pages 9 – 15

Page 9

$\frac{1}{3}$ $\frac{1}{4}$: 3, 6, 9, **12**; 4, 8, **12**

$\frac{1}{6}$ $\frac{1}{8}$: 6, 12, 18, **24**; 8, 16, **24**

$\frac{1}{4}$ $\frac{1}{7}$: 4, 8, 12, 16, 20, 24, **28**; 7, 14, 21, **28**

$\frac{1}{5}$ $\frac{1}{9}$: 5, 10, 15, 20, 25, 30, 35, 40, **45**; 9, 18, 27, 36, **45**

$\frac{1}{7}$ $\frac{1}{6}$: 7, 14, 21, 28, 35, **42**; 6, 12, 18, 24, 30, 36, **42**

$\frac{1}{4}$ $\frac{1}{10}$: 4, 8, 12, 16, **20**; 10, **20**

$\frac{1}{8}$ $\frac{1}{12}$: 8, 16, **24**; 12, **24**

$\frac{1}{9}$ $\frac{1}{6}$: 9, **18**; 6, 12, **18**

Page 10

$\frac{7}{10}$ $\frac{4}{6}$ or $\frac{2}{3}$

$\frac{4}{12}$ or $\frac{1}{3}$ $\frac{4}{9}$

$\frac{9}{16}$ $\frac{8}{15}$

$\frac{13}{18}$ $\frac{7}{9}$

$\frac{8}{9}$ $\frac{13}{14}$

$\frac{6}{12}$ or $\frac{1}{2}$ $\frac{3}{8}$

Page 11

$\frac{1}{2}$ $1\frac{5}{12}$

$\frac{3}{4}$ $\frac{5}{8}$

$\frac{5}{8}$ $1\frac{1}{10}$

$1\frac{1}{6}$ $1\frac{5}{42}$

$\frac{13}{21}$ $\frac{35}{72}$

$1\frac{19}{70}$ $1\frac{5}{24}$

$\frac{17}{25}$ $\frac{43}{56}$

$\frac{11}{30}$ $1\frac{11}{35}$

$\frac{11}{12}$ $\frac{7}{10}$

$\frac{27}{40}$ $1\frac{8}{63}$

Page 12

$\frac{1}{6}$ $\frac{4}{15}$

$\frac{5}{12}$ $\frac{1}{24}$

$\frac{2}{9}$ $\frac{4}{15}$

$\frac{7}{20}$ $\frac{7}{12}$

$\frac{3}{8}$ $\frac{1}{6}$

$\frac{5}{9}$ $\frac{7}{18}$

Page 13

$\frac{1}{5}$ $\frac{1}{12}$

$\frac{11}{16}$ $\frac{3}{8}$

$\frac{5}{24}$ $\frac{1}{10}$

$\frac{1}{6}$ $\frac{1}{2}$

$\frac{2}{9}$ $\frac{5}{54}$

$\frac{11}{70}$ $\frac{13}{24}$

$\frac{13}{25}$ $\frac{27}{56}$

$\frac{7}{34}$ $\frac{4}{35}$

$\frac{5}{32}$ $\frac{1}{10}$

$\frac{3}{40}$ $\frac{13}{63}$

Page 14

$\frac{1}{6}$ $\frac{1}{5}$

$\frac{1}{6}$ $\frac{2}{9}$

$\frac{5}{24}$ $\frac{1}{3}$

$\frac{3}{14}$ $\frac{1}{2}$

$\frac{4}{27}$ $\frac{3}{28}$

$\frac{5}{18}$ $\frac{1}{54}$

Page 15

$\frac{5}{9}$ $\frac{1}{9}$

$\frac{7}{12}$ $\frac{1}{12}$

$\frac{2}{21}$ $\frac{1}{18}$

$\frac{1}{4}$ $\frac{1}{2}$

$\frac{1}{9}$ $\frac{1}{5}$

$\frac{3}{16}$ $\frac{1}{18}$

Fractions II—Operations   © Edupress EP160

Advanced Fractions

Answers Pages 16 - 22

Page 16

$\frac{1}{2}$	$\frac{1}{72}$
$\frac{7}{18}$	$\frac{1}{3}$
$\frac{1}{4}$	$\frac{10}{21}$
$\frac{1}{2}$	$\frac{1}{2}$
$\frac{45}{128}$	$\frac{1}{4}$
$\frac{4}{9}$	$\frac{8}{15}$
$\frac{1}{9}$	$\frac{121}{420}$
$\frac{1}{4}$	$\frac{1}{7}$
$\frac{1}{4}$	$\frac{1}{9}$
$\frac{1}{15}$	$\frac{2}{7}$

Page 17

$\frac{3}{4}$	$1\frac{1}{3}$
$\frac{3}{5}$	$\frac{1}{6}$
$6\frac{2}{3}$	$\frac{12}{35}$
4	$\frac{11}{36}$
3	$1\frac{5}{16}$

Page 18

$\frac{8}{9}$	$\frac{15}{16}$
$1\frac{1}{3}$	$\frac{16}{27}$
$1\frac{11}{25}$	$\frac{24}{25}$
$\frac{20}{27}$	$1\frac{1}{2}$
$1\frac{7}{128}$	$1\frac{1}{15}$
$\frac{5}{7}$	$\frac{27}{64}$
$\frac{1}{3}$	$\frac{9}{14}$
$1\frac{13}{36}$	$1\frac{19}{35}$
$1\frac{17}{18}$	$\frac{2}{3}$
$\frac{2}{5}$	$\frac{9}{14}$

Page 19

$\frac{13}{24}$	$\frac{3}{8}$
$\frac{1}{9}$	$\frac{25}{36}$
$\frac{1}{3}$	$1\frac{2}{15}$
$1\frac{7}{20}$	$\frac{1}{21}$
$\frac{11}{16}$	$\frac{15}{16}$
$\frac{1}{15}$	$\frac{1}{3}$
$\frac{24}{49}$	$1\frac{29}{63}$
$\frac{1}{4}$	$\frac{11}{21}$
$\frac{17}{42}$	5
$1\frac{1}{5}$	$\frac{2}{7}$

Page 20

$1\frac{2}{15}$	$\frac{2}{7}$
$\frac{27}{40}$	$\frac{1}{2}$
$\frac{2}{5}$	$1\frac{25}{66}$
$\frac{4}{5}$	$\frac{3}{26}$
$1\frac{5}{24}$	$\frac{3}{4}$
$\frac{1}{63}$	$\frac{22}{27}$
$1\frac{1}{48}$	$\frac{35}{72}$
$\frac{1}{9}$	$\frac{19}{66}$
$\frac{11}{70}$	$2\frac{1}{12}$
$1\frac{5}{9}$	$\frac{3}{4}$

Page 21

6	$6\frac{3}{4}$
6	4
40	$5\frac{5}{9}$
18	$4\frac{3}{8}$
$2\frac{8}{11}$	2
$5\frac{1}{4}$	3

Page 22

$7\frac{7}{20}$	$7\frac{8}{15}$	$11\frac{2}{3}$
$\frac{1}{2}$	$5\frac{13}{42}$	$2\frac{7}{15}$

Fractions II—Operations

© Edupress EP160

Advanced Fractions

Answers Pages 23 - 28

Page 23

$8 \frac{1}{4}$

$2 \frac{4}{15}$

$28 \frac{1}{18}$

$1 \frac{135}{161}$

$24 \frac{7}{24}$

$1 \frac{8}{25}$

$9 \frac{11}{21}$

$\frac{14}{15}$

Page 24

6

$1 \frac{2}{3}$

$1 \frac{19}{30}$

$10 \frac{5}{28}$

$1 \frac{9}{10}$

$11 \frac{47}{132}$

$1 \frac{19}{91}$

$8 \frac{3}{4}$

$10 \frac{1}{6}$

$3 \frac{16}{35}$

$2 \frac{3}{4}$

Page 25

$1 \frac{5}{28}$ $\frac{5}{7}$

$\frac{1}{5}$ $1 \frac{1}{8}$

$\frac{6}{35}$ $\frac{31}{36}$

$\frac{4}{7}$ $\frac{13}{25}$

$1 \frac{5}{56}$ $\frac{5}{6}$

$\frac{2}{15}$ $\frac{1}{6}$

$10 \frac{1}{6}$

$1 \frac{13}{15}$

$8 \frac{11}{14}$

$2 \frac{6}{7}$

$4 \frac{11}{13}$

Page 26

$1 \frac{4}{15}$ $\frac{9}{25}$

$\frac{23}{35}$ $\frac{4}{5}$

$\frac{9}{35}$ $1 \frac{17}{60}$

$3 \frac{15}{44}$ $\frac{18}{35}$

$\frac{26}{55}$ $2 \frac{4}{7}$

$\frac{8}{35}$ $\frac{60}{221}$

$9 \frac{31}{35}$

$15 \frac{31}{54}$

$9 \frac{13}{24}$

$6 \frac{2}{5}$

$2 \frac{1}{2}$

Page 27

$\frac{5}{7}$ $\frac{3}{4}$

$\frac{2}{9}$ $\frac{2}{5}$

$1 \frac{5}{72}$ $1 \frac{5}{24}$

$\frac{47}{100}$ $1 \frac{1}{48}$

$\frac{3}{16}$ $\frac{2}{9}$

$\frac{2}{15}$ $\frac{15}{28}$

$\frac{1}{2}$ $\frac{32}{63}$

$\frac{18}{29}$ $\frac{5}{12}$

2 $\frac{1}{2}$

2 $\frac{1}{6}$

Page 28

$3 \frac{3}{5}$

$14 \frac{2}{3}$

$6 \frac{6}{7}$

$7 \frac{2}{13}$

$10 \frac{7}{18}$

$7 \frac{41}{60}$

$1 \frac{11}{12}$

$25 \frac{5}{12}$

$18 \frac{35}{48}$

$2 \frac{1}{10}$

$2 \frac{3}{4}$

Fractions II—Operations   © Edupress EP160